Dorset Sheep, or; The Winter Lamb

by J.E. Wing

with an introduction by Jackson Chambers

Self Reliance Books

Get more historic titles on animal and stock breeding, gardening and old fashioned skills by visiting us at:

http://selfreliancebooks.blogspot.com/

Introduction

I am pleased to present yet another practical title on breeding and raising livestock.

The work is in the Public Domain and is re-printed here in accordance with Federal Laws.

As with all reprinted books of this age that are intended to perfectly reproduce the original edition, considerable pains and effort had to be undertaken to correct fading and sometimes outright damage to existing proofs of this title. At times, this task is quite monumental, requiring an almost total "rebuilding" of some pages from digital proofs of multiple copies. Despite this, imperfections still sometimes exist in the final proof and may detract from the visual appearance of the text.

I hope you enjoy reading this book as much as I enjoyed making it available to readers again.

Jackson Chambers

The Celebrated Blackfaced Ram "CAIRNTABLE,"
Bred by CHARLES HOWATSON, of Dornel, and sold for SIXTY POUNDS,
To J. N. FLEMING, of Keil   July, 1870

PREFACE

It has seemed to the authors that, with all the long list of works on sheep, nothing quite met the case of the Dorset Sheep under American conditions. This work is the result of their experience. They have purposely given prominence to things that in their experience have proved important and slighted others that have not seemed of much importance in their management. Failures with Dorset sheep are usually the result of ignorance. Methods that will result fairly well with some breeds, for instance with Merinos, will not do for Dorsets if one wishes to get from them their peculiar possibilities in the way of profits. Finally we submit this as an effort towards help and feel that the breeder must be a past master in the gentle art of shepherding not to find some hints here that will be useful to him. And we crave your kind indulgence if we have slighted some things and magnified others; we have spent the most time telling of the turns where we ourselves missed the right turnings.

H. H. MILLER.

JOSEPH E. WING.

THE DORSET EWE

COUNTY DORSET is one of the most southerly of the counties of England. It is a warm, sunny, sheltered county, having hills and valleys, arable lands and pastures. It is one of the oldest civilized parts of England. Here the Romans landed; here they builded cities and walled them; today some of the walls are standing, and the roads are often as laid off by them.

When the breed of Horn sheep was first introduced into Dorsetshire history does not recount. Doubtless the breed was there in Roman days. It has been there ever since. No other breed has been able to supplant it, though at different times other sheep have been popular in parts of Dorsetshire for a time. They have always had to give way, however, to the old breed of Dorset Horns. Why is this?

THE DAIRY SHEEP.

Go back half a century and more. Sheep were the dairy animals of Dorset, and parts of the adjacent counties of Somerset and Devonshire. The best milking ewes were selected, their ewe lambs were retained. Already the Horns were famous for their milking qualities. This use intensified the qualities. From this time and this custom then comes the wonderful milking properties of the Dorset ewe. "I was raised on Dorset milk," remarked Dick Stone to the writer. This milking trait, once so well established, formed the foundation for the somewhat later development of the early fat lamb business.

THE EARLY LAMB.

Sheep in a state of nature drop their lambs on grass, and it is hard to induce them to conceive to drop them earlier. The Dorset is so much an artificial production, has been so long under the moulding hand of its shepherds, that it has lost this instinct and now the lambs come in the late Fall or early Winter months. The shepherds have provided abundant winter food for so long the ewes have learned to look for it; the habit has been so firmly fixed that Dorset ewes habitually lamb in November, December and January. And they will lamb earlier than that if rightly treated. This habit is of the utmost importance and value. It is this habit that gives the Dorset ewe her great value in her native land. There her lambs are born out of doors, preferably in the pretty, wind-protected valleys—it is a mild clime, in these sheltered valleys—there is generally some grass, there are turnips and hay, and the little lambs are fed with their mothers upon our corn and what they term "cake," that is oilmeal. The little lambs grow prodigiously, filled to bursting by their mother's milk, and eating grain like little pigs. Lambs in Dorset will even get too fat to walk, as little pigs will sometimes in our land. In winter and early spring these round, plump, sweet, juicy lumps of baby mutton go to the London butchers. They bring good prices, what would be considered fabulous prices in our country.

Thus it is that the shepherds of Dorset stick by their Horn sheep. It is, perhaps, first a sentiment, it is next a matter of profit. No other sheep in the world has in it the capacity for profit that the Dorset Horn has. And this is true in America as well as in her native county of Dorset.

GETTING FALL LAMBS.

In taking the Dorset away from her native hills it must be borne in mind that you have changed the conditions materially. To get the same results that have been gotten in Dorsetshire then you must in some measure supply similar conditions. How are we to do this? Well, let us start from what we might call a basis of comparison. In Dorsetshire the ram is usually coupled with the ewes during June and July, but in this climate (referring to the Eastern and

YEARLING RAMS

Middle States) that any large and uniform success in breeding Dorsets as late as June and July will result, we think improbable. Why? Because it is too hot then. Now most seeds require great warmth to germinate; some, like the sweet pea, prefer cool conditions, so we plant them early. The same reasoning applies to sheep. Their natural time of mating is fall, October and November, cool months. So if we want them to breed in the spring we should select cool periods. This seems a simple thing. Yes, it is. And like many simple things is overlooked. When we started with Dorsets we were told and read they would breed any time of the year; also, that June was the month to mate them for fall lambs. We tried them in June for several seasons, but with partial success only. This experience convinced us that while Dorsets will, in isolated cases, breed any time of the year, that for uniform and complete success the "any time" must be a time when conditions are right. We were now on the true track and realized that for spring breeding we must select a time as near like the natural period of fall as possible, and (equally important) have the ewes as near like their natural fall condition as possible. To meet these desired conditions we suggest the observance of the following:

1st.—Have the ram with ewes not earlier than middle of March, not later than middle of May.

2d.—Put ram with ewes nights, not days.

3rd.—Use young ram, and feed him well while in service.

4th.—Do not have ram too fat.

5th.—Do not have ewes too thin.

6th.—If ewes were not shorn in fall, shear as early as you dare.

7th.—Feed ewes green food, such as ensilage, turnips, carrots, mangels, etc., with some corn.

If ewes lambed previous fall and are dry, this feed is not necessary until a few weeks before you are to put the ram with them. But no matter what the condition of ewes, no matter what you have been feeding, increase the quantity at this time. For remember that during fall a ewe is naturally gaining flesh, while during spring the tendency is to lose flesh. Reverse this tendency as far as a little feed will do it, and make that extra food as nearly "green as grass" as you can. And remember, never, NEVER let lambs be born later than April or before September. Separate your rams from your ewes the first of November. Any ewes not with lamb then will breed for next fall and late lambs are of little value as a rule and to be discountenanced among Dorsets.

GETTING THE PROFITS.

I have mentioned the feeling of sentiment that doubtless contributes to the persistency with which the shepherds of Dorsetshire cling to their favorite breed. In our country there is little of sentiment in sheep breeding; we are newer at the business; we have an immense expanse of land; a varied climate, and nearly all breeds of sheep in all parts of the country. There is none of that confining of a certain breed to a certain county, or even state or section. And it is this faith in the breeding of one kind of sheep by the inhabitants of a limited area that creates sentiment. But of profit—well, if all haven't it, all want it. And I say again, no breed has a greater capacity for profit than the Dorset. Properly understood, and consistently handled on the basis of such understanding, a small flock will surely add a pleasing sum to the annual income, while a good sized flock, such as the average farm of the Eastern and Middle States could surely carry, will represent a good living. For instance, I know of one man who for years has realized about $2,000 annually from a flock of about 100 ewes. This is equal to the sales from a good sized dairy, yet the cost of feed and labor would be much less. Surely as an early lamb raiser the Dorset is a money maker. When you have studied the breed and business, make a start, you will then realize fully the pleasures and profits.

RYE AND OATS PASTURE.

Here is a combination that will furnish most excellent fall pasture. And how the ewes and lambs will relish it! Winter rye is often pastured in the spring; but with oats, sown as a catch crop on odd bit of ground, it is especially certain and cheap as a fall pasture. Unlike rape, rye will grow on poor land, while oats in the fall will grow wherever they're dropped. Everyone can find a place for a few acres.

If no other available land, after corn is cut, run cultivator or riding harrow over corn stubble and sow broadcast. If your corn is cut clean, not too many ears lying on the ground, you can turn in few hours each day before husking; however this practice requires watchfulness, for while it will take the lambs a long time to discover and eat corn ears, the ewes will soon nose them out and if left too long at a time may eat too many. When they do get to eating them, if you leave them all day, you'll very likely find a dead ewe or so at night. Some may say, why not pasture a regular crop of rye and not bother with this extra piece? Well,

you can; we have ourselves, but the practice has many objections. The crop may not be much injured if it has good top, the young timothy (usually sown with winter rye) will though, and then the ground will be compacted so as not to be in as good condition for clover seeding the following spring. It is much better to have a piece by itself, you'll get two or three times the amount of feed from it and won't worry over possible injury. Just make up your mind to have it and you'll find the land. Sow any time during September, use twice as much rye as oats, put on thick, not less than three bushels to the acre. When up so as to look green and grassy turn on, don't be afraid, as there is no more risk than with grass. It is good practice though, with any special fall pasture, to turn on grass first for a few hours. Your other pasture will last longer, and your sheep will do better.

SUMMER CARE OF PREGNANT EWES.

Good summer care is a very simple matter. Three things are necessary, grass, water, shade. In the matter of grass, either good native pasture or a run in clover or alfalfa will do. There is this danger of clover or alfalfa pasture for the ewe not giving suck to a lamb, she is apt to become too fat. There is nothing equal to blue grass at this period. But in using it one should remember the danger of parasitic infection that comes from the use of blue grass. We will speak of this later. Water should be clean, as fresh as possible to supply, kept in raised troughs that can not become fouled with excrements. There is danger of parasitic infection from the drinking water. This is especially true when the sheep are required or allowed to suck up their drink from marshy seeps, tiny streams, grass-bordered or stagnant pools. The excrements roll down and pollute the water, the germs that they doubtless contain are hatched there, the sheep take them in again and in ever increasing numbers. And while mature ewes will not often pine away and die as lambs will, yet none the less is it weakening to them to be preyed upon by these internal parasites. Not only stomach-worms but tape-worms are spread by the too continual use of pastures. It is a safe rule never to stock a pasture to its capacity with sheep. Better always keep half on it what it would carry and graze it also with cows, or perhaps horses.

SHADE.

You may depend upon it that your flock will seek shade and will need it. If you are wise you will train them to come to the barn, or to sheds, where they will get the most comfort and their droppings will be under cover. There are

two things gained by this. The manurial value of the droppings is saved, whereas if they are piled year after year
beneath forest trees they do you very little good, and the

SHEEP BARN, 36x36, ON WOODLAND FARM.—NOTE THE AIRINESS.

danger of infecting the land is much reduced. I know of
fields where ewes always lie along the old fences seeking
shade. There is a strip of very rich, rank grass along

these fences. This grass is deadly to lambs and dangerous to ewes because of the parasites that it harbors. The lamb is often hungry. He lies down a little while, then jumps up, goes a little way and nibbles the grass. He takes in the germs and perishes from them in due time. This, if there was no other reason, would lead me to condemn the practice of allowing ewes to shade along fences.

PUMPKIN FEEDING.

As fall comes on the grass is perhaps short and dry and there is room for some artificial feeding. Of all the substances that may be grown and fed to ewes and lambs in the fall none compare with pumpkins. First, they may be grown so cheaply. Our practice is to grow them in the corn where they are a catch-crop of almost clear profit. We find it necessary to plant a great many seeds in order to get what pumpkins we need, owing to the ravages of the small striped bug, and this is the only secret we have ever observed in getting a crop of pumpkins. Certainly, they need rich soil. In feeding we never remove the seeds as they are the richest parts and the part that give to pumpkins their great value to the shepherd. Pumpkin seeds are among the best vermifuges known. They destroy and expel tape worms, and I think clear out many forms of worms; certainly I do know that you may take an old ewe, her skin white, her eye dull, every appearance of her being diseased, feed her all the pumpkins she will eat for a few weeks and she will renew her youth. When we have them in abundance we haul them out by the ton, simply strewing them about the pastures and allowing the sheep to gnaw into them at their pleasure.

CARE IN PUMPKIN FEEDING.

Like many another good thing, however, pumpkin feeding may be carried too far. There is danger that the pregnant ewe may become too fat, if allowed all the pumpkins she will eat. This is the only danger and the remedy is easy.

PUMPKINS FOR MILKING EWES.

After the little lambs come is the time when pumpkins come in good play. There is nothing so good for the milking ewe. Soon, too, the little lamb will begin to nibble them. They will do him no harm, though he will need stronger food with them, grain of some sort. These pumpkins will keep indefinitely, and may be used until about Christmas. I give this much space to the pumpkin because it is of such easy culture and surely no farmer who is keeping sheep need try to do without this help.

THE RAPE PLANT.

Quite often rape will be a great help. If a small field can be sown early to tide over the hot, dry part of summer and fall it will be of great use, and it may be sown in the corn at the time of last cultivation where it will to a large extent keep down weeds and after the corn is cut it will come on if the season is favorable, and make a lot of fall and early winter pasturage. It will be necessary to haul away the corn before sheep can be turned in, as they soon learn to hunt for the ears and gorge themselves. In truth in cutting the corn, or in husking it from the stalk, unusual care should be taken not to let too much corn remain scattered on the ground. Dorset ewes have the sharpest eyes and the best appetites of any sheep and will glean every ear before they do much else. There is danger, of course, of their getting too much corn. Rape is generally safe feed, though there are times when it will bloat the ewes. It is not usually safe to allow them to remain constantly on it for they will become

too fat. This is not true after lambing. We have seen
them bloat to distress on rape but have never had any die,
and there is probably little danger from feeding it at any
time. It should never be turned on when frozen, not that
it will hurt the ewes so much but each leaf that is bent or
disturbed when frozen, will be killed and wasted. It takes
a cold of about 12 degrees to kill rape. In selecting a field
for rape good land should be chosen and it should be remembered that ewes will need to run thereon when it is moist
so that if it is a field of clay, apt to pack hard, it may be
unwise to sow to rape. If the sheep are taken off at Christmas, however, there will generally be time for frost to liven
up the packed land.

"Rape may be sown with oats, barley, winter rye or
wheat. If sown with winter rye or wheat, harrow the rye
field in early spring and sow about two pounds of rape
seed per acre, harrowing lightly again after the seed has
been sown. Such harrowing will usually be helpful to the
rye crop. Rape seed can be sown with oats or barley, but
if this is done the growth of rape is liable to become so rank,
especially if the season is a wet one, that the plants will
grow as tall as the oats or barley. When this happens trouble occurs at harvest time owing to the green rape plants
being cut and bound in the sheaves, causing them to rot
under the bands. The following is a better plan: Eight or
nine days after sowing the oats or barley, when the young
grain plants are three or four inches high, run a slant-tooth
harrow over the field to loosen the soil. Then seed two or
three pounds of rape and harrow lightly again. By seeding in
this way the grain crop has so much the start of the rape that
the latter is kept small and spindling until grain is harvested.
After harvest the rape plants getting the benefits of sun and
moisture begin to grow, and in good season the field will
soon be covered with green forage. Rape seed can be sown
broadcast any.time from April until August. For broadcast

seeding prepare the land as for oats and sow three or four pounds of seed per acre and harrow in lightly. Land on which rape is sown broadcast should be comparatively free from weed seeds and in good condition generally."—HENRY.

CARE OF FALL AND WINTER LAMBS.

Fall lambs come strong and can look after themselves; there is little or no trouble with the ewes, the most important

A GOOD TYPE

thing is to watch their udders, for they have full ones at that time. We make a practice of putting the ewes on thin pasture a few weeks before the lambs are due, then as fast as they drop the lambs we take them to the barn, where they are kept for a few days, getting a bite of hay and a bit of grain. As soon as the lambs take all the milk, the ewes can be put on full pasture of grass, rye and rape, until snow flies. One thing to bear in mind is, these lambs should never get wet; fall storms are not like summer showers, and they are very severe on young lambs, even the heavy dews of fall nights where the pasture is tall and heavy should be avoided.

For this reason and also because it is safer for the ewes, pasture the rye, rape and new seeding of grass during the day, and put them on the short grass at night. The best way of all is to bring them to the barn for over night; they will have shelter in case of sudden storm, and anyway when the lambs are a few weeks old you will want to teach them to eat grain; the barn is the place to do this. They should have a room or space all to themselves; it should be shut off from the main building by a door that slides up and down. This door should have a space for a creep that can be opened and shut independent of the door. When all the lambs are in their room, shut the door and make them stay there until they have eaten all their feed. The creep can then be opened so they can run in and out to nibble at the hay. This method is much more effective in results obtained than using the creep alone. With the latter only many of the lambs, especially the younger ones, will spend most of their time creeping in and out, while the others eat all the feed or muss what they do not eat. Another advantage with Dorsets is that fewer horns are broken, the lambs never rushing and crowding through the creep. It is very easy to teach the lambs to go into their own room. At first you may have to catch a few, but they will soon learn to run right in, one following the other.

The time the ewes and lambs can run on pasture will of course vary with the seasons. But as long as they are on pasture one feed of grain per day will be sufficient. And the way the lambs will grow with the pasture and the one feed will be a revelation to all who have only handled spring lambs.

GRAIN FEEDING BEFORE LAMBING.

If ewes are in good heart it is never necessary to feed grain before lambing unless in small amounts. It is not often safe to feed much grain to the pregnant Dorset ewe.

A CONTINENTAL FLOCK. (COURTESY CONTINENTAL DORSET CLUB RECORD.)

18

The result of too much grain feeding is apt to be a weak lamb, hard to induce to live, whereas Dorset lambs are, when their mothers are rightly managed, the strongest lambs in the world. Dorsets are hungry sheep. They will always, if not sick eat everything before them. There is no sheep with a better appetite or digestion. The inexperienced shepherd is quite apt to over-feed them. Good, sound clover or alfalfa hay is quite good enough for the pregnant ewe after green stuffs are gone. Let her have a plenty of it. If you must feed some grain to keep her in flesh because of the badness of your hay, feed oats and bran, equal parts by weight. There is no sheep easier kept in flesh if she is not worm-infested. If she is cared for as she should be she will not be that.

HOUSING.

It is not well to keep the pregnant ewe very closely housed. She ought to have a good run and every day when it is not actually storming she should be out. Sometimes the run of a dry feed-lot, with coarse fodders to pick over in the yard will be sufficient, and this course has the advantage of the flock being always in view and stray lambs being born are apt to be seen. With others a bit of grass of ten acres or more, not too closely grazed in the fall, will be provided and on this the ewes will take a great deal of pleasure and get quite a good deal of nourishment. They must at all times be in the mind of their shepherd, for lambs may be born out on the grass or in the snow, but you must not let this fear deter you from giving them their daily run out of doors. It must not be thought, however, that because a certain amount of outdoor life is good no shelter at all is better. A comfortable barn is needed, and, in truth, in the Northern states indispensable.

SHEDDING.

A comfortable shelter, closed tight on the north, west and east sides, with chance to open well on the south and preferably with considerable glass where it will let in the sun, is what you need. It need not be an expensive structure. It is better to have storage for hay above. There must be ample provision for fresh air to come in from the south, so that cold blasts will not come with it. Let there be a yard attached, preferably on the sunny side. Water may be in the barn, or in the yard unless in a very cold country. Every night the flock should be confined to their barn. It will be found that the lambs will mostly come in the night. And if the doors are open it will generally be found that the ewes will come to the barn to drop their lambs. Too close shutting in will work harm to the flock. Too much exposure will cost you their thrift and the loss of some lambs in severe weather. In the South, Dorsets thrive with no shelter at all save that afforded by hill, tree and shrub. Yet, in general, it will be found that it will take less feed and the flock will keep in better condition to shelter them especially from all rains in cold weather.

WINTER FEEDS—ENSILAGE.

There is great diversity of opinion as to the value of ensilage for sheep, or rather to be more accurate, the difference of opinion is more as to whether it can be fed with safety, for the ensilage itself is generally admitted to be a good feed. Some sheep feeders will not use it at all; some of our experiment stations condemn it as a sheep feed. For our part, we have fed it for many years to both lambs and ewes, and consider it both safe and of much value, especially so for ewes with lambs by side. But it must be used with judgment, which means not to feed too much or too often. Our practice is to feed once a day during winter or cold weather only; we take daily from the silo the amount of a day's feed, put it by itself and let it remain for a few hours until it becomes cool to the touch. Late in the spring or during hot weather we would not feed it to sheep. Also, let sheep feeders beware of it when taken from near the bottom of the silo; it is then very wet and chuck full of acidity, a slow poison for sheep. You don't need a silo in the sheep business, but you can use one if you have it and want to. If you keep Dorsets they will thrive without ensilage, so will you.

ROOTS.

Sheep without roots! Hamlet without the Ghost! Carrots, turnips, mangels, there is no question about the value and safety of this trio. I name them in the order of their excellence as sheep feed. I might add that carrots are the hardest to grow, the best to feed; turnips the easiest to grow, the most universally fed; mangels the surest to grow, the poorest to feed. They all need rich ground, all will do better on a rather heavy soil, but you can get a good crop from light soil well enriched. Carrots should be sown about corn plainting time in rows two and a half feet apart, the rows slightly ridged, this ridging facilitates weeding, makes easier pulling; sow quite thick to insure good stand, as they are shy starters; when up a few inches thin with a broad hoe, leaving little bunches between strokes of hoe; thin these bunches by hand to one plant. After this there is little work. Use the large stock varieties, not the table carrot. Turnips can be sown from middle of June to middle of July in rows two and one-half or three feet apart, either flat or ridged high; sow them thin but even; they are quick, easy starters; when well up, thin at once (don't let 'em get big) with hoe to one plant twelve or fifteen inches apart. An occasional cultivation afterwards is all that is needed. Use the Swede varieties. Mangels should be sown early in the spring; sow same as turnips, only thicker, as they are slow to start and many seeds will not sprout if a bit dry; care for them same as turnips, but thin farther apart. The ewes and especially

the lambs, will be crazy for the carrots, and you are not apt
to have enough to feed too many.   They will eat the turnips
eagerly, too, and many shepherds think you can't feed them
too many; we have had large experience in feeding turnips,
having used them in unlimited quantities for ewes, rams and
lambs.   We think pregnant ewes can be fed too many, and
that it is better to limit them to one moderate feed per day
until after lambing, when they can safely have as many as

they'll eat.   Mangels contain the most water of the three,
and are rather chilly eating on a cold winter's day; they are
perhaps the best keepers, though, and are very acceptable to
sheep in late spring, although we have the idea they do not
like them as well as turnips, and know that carrots are "pie"
to them compared with mangels. Care must be used in feeding mangels to rams, as in quantity they have deleterious
effect upon the bladder.   If you keep Dorsets, grow some
kind of roots.   Dorsets are the alchemists among sheep, and
will turn them into gold for you.

CLOVERS AND LEGUMES.

You are engaged, now, in making milk and baby flesh. Each is largely composed of protein, to produce which you must feed protein. That means to buy large amounts of wheat bran, gluten feed or oil meal, or it means to produce your own protein supply. You can do this most easily by growing red clover, alfalfa, soy beans or cow peas.

ALFALFA.

This is the richest and most easily grown hay in the world. Sheep love it. It is the best maintenance ration for ewes before lambing in winter and the best basis for any combination of feeds for them after lambing. And almost anyone can grow it who has sheep. The reason is that it requires, MUST have, rich soil, and sheep make manure that will enrich that soil. Take an acre or two as a beginning, on dry, pervious soil, where it is dry and firm in winter, apply manure liberally, plow deep in early spring, work down to a good tilth and sow one or two bushels of spring beardless barley to the acre and 15 pounds of alfalfa seed. Cover the seed lightly. Roll it if not too moist. Cut the barley for hay or grain and mow the alfalfa close once or twice that summer after the barley is taken off. Keep all stock off during cold or wet weather; in fact, keep them off at all times for the first two years. You will now have a set of alfalfa that will give you from three to eight tons per acre the second year and for many years thereafter. Mow the alfalfa as soon as bloom appears in the spring and at intervals of about thirty-five days thereafter. And sow another acre or two as you get the manure and the experience. Woodland Farm began ten years ago with an acre and now cuts nearly 250 tons yearly besides pasturing a good deal.

There is absolutely no danger in feeding alfalfa hay, but

DORSET EWES—RIGHT TYPE

24

there is need of care in pasturing green alfalfa. The danger comes from bloat and that is the result of indigestion, caused by the animals eating too greedily of the delicious green feed. We let the alfalfa grow up about twelve inches tall, then, when the sheep are full of green grass, and at about ten o'clock, when the sun is warm, turn them into the alfalfa. They remain there constantly except that they come to the barn to shade during the heat of the day. They go back as soon as they care to graze. Treated in this way, we have had no loss from bloat, but have had magnificent results in development of our young sheep. CAUTION.—After frost alfalfa should never be grazed, as it is apt to cause indigestion and death. There is no crop that will return so much feeding value per acre as alfalfa, if you are on alfalfa land, or will take the trouble to make your land alfalfa land. If your land is deficient in lime it should be well limed AFTER it has been well manured; lime alone will not make alfalfa land of it. If it is based on shale, close to the surface, it may not pay to sow alfalfa, which roots quite deep.

RED CLOVER.

What will secure a stand of alfalfa will also secure a stand of red clover, though, as it is a biennial, it need not be expected to remain in the soil more than two seasons. It has a rich nutritive value and should always be cut for sheep before the heads are brown.

SOY BEANS.

In the Northern states the soy bean is a rich gift. Planted in drills about 22 inches apart, cultivated once or twice, cut with a binder and threshed, they yield from 15 to 30 bushels per acre of extremely rich beans, which will go far towards balancing a ration. Sheep are very fond of these beans, and also of the leaves and stems, if saved without rain.   John B. Peelle, who is a leading hot-house Dorset lamb producer relies greatly on his soys and saves immensely in his feed bill

READY FOR NEW YORK AND GOOD FOR $10.00
(From Peelle's Place)

thereby, besides producing $10.00 lambs in abundance.   Sow early varieties north.

COW PEAS.

In the Southern states and along the border there are regions where neither red clover nor alfalfa are to be depended upon and where soils need building up to fit them for other crops.   Here the cow peas come in exceedingly well. They are great soil builders, rich in protein, make good summer or winter forage and are the great reliance of the Southern shepherd. Cow-peas sown among ensilage corn help to balance that ration; sown among soiling crops of rape or sorghum serve to balance them and enrich the soil at the same time. They must not be sown until after the land is warm in the spring.

CANADA FIELD PEAS.

There is sometimes a confusion of terms. The Canadian field peas are much like ordinary garden peas, and must be sown very early, on good soil, with oats or barley as a soiling or pasture crop. Cow-peas are really beans, must be sown late, will grow on soil that will not produce Canada peas. The Canada peas and barley make an ideal soiling feed, or the lambs may run through creeps and eat the mixture, and will thrive thereon first rate. The advent of hot, dry weather finishes the Canada peas, however.

PUTTING THE LAMBS FORWARD.

A good old English practice is to hurdle the field with creeps so that the lambs may "run forward" of their mothers, thus getting the first bite of the fresh feed. From time to time the hurdles, or panels of movable fence, are moved up and the lambs' ewes clean up what the lambs have been over. This is a good way to make fine lambs at small expense and to keep them free from parasites. The best of all for the babies always.

SHEARING.

There is no sheep easier to shear and shear well than the Dorset. The wool cuts easily; the operator can as well as not leave a smooth, close-cut surface. The machines work well on Dorset sheep, and some of the best American flocks are shorn by machines altogether. The use of the machines is most discouraging to ticks, which are almost certainly cut in two and destroyed. Care should be taken not to shear too close after flies are troublesome, as there is not enough wool to protect the sheep after the machine shear has been over her back. The last week in March or early in April is a good time to shear the flock; in warm climates it is often well to shear again in late August. This double shearing does not make much more wool, perhaps no more at all, but it is a relief to the overburdened sheep.

MARKING.

The English method of marking by branding figures in the horn is a good one but slow, and necessarily the horn must first grow so that some means of marking the lambs must be adopted. Ear labels of various sorts are on the market, and all are open to the objection that they lose out. Some breeders use the tattoo mark with success, though others fail in using it. The secret of success with the tattoo mark is, first, see that the jaws are exactly parallel. They may be made so by careful use of the file. The points of the letters should indent evenly a thick piece of paper. Next, plenty of India ink should be used. The points should be firmly pressed in and immediately the ink must be rubbed into the wounds. The advantage of the tattoo mark is that it does not deface the ear, is absolutely permanent and can not be transferred by any trickster from one sheep to another. Tattoo markers are made by F. S. Burch, 178 Michigan Street, Chicago, Ill. The first cost is rather heavy, but in the long run there is a saving, as the ink is cheap.

WEANING.

Don't be in a hurry to wean lambs that are to be kept. There is nothing like mother's milk, unless it is more mother's milk! Let the lambs have access to all the bran and oats they can eat; all the green feed and the mother's milk, too. You can shorten the time of development at least one year by liberal feeding. It takes less feed to make a sheep if it is fed in one year than if it had been fed in two

years. When the lambs are separated from their mothers take away the ewes; the lambs fret very little. If there is yet milk in the ewes remove it a few times, not quite clean.

DIPPING.

"A man ought to bathe once a year, whether he needs it or not." So of the sheep. Dip them once a year, whether they need it or not. There are almost surely a few ticks, maybe a few lice on them. Dipping costs but a trifle. Provide a steel tank, galvanized. Sink it in the floor of your sheep house. When not in use cover it with good planking. A tank six or eight feet long will answer for a small flock, and as they are all narrow it takes but little stuff to fill them. Have the draining pen long and narrow, so that as the sheep walk up one at a time they may be let out ahead. Pen with movable hurdles or panels. Half a day with three active men will dip a flock of a hundred. The carbolic dips are safe and good. There is no profit in ticks, though there is much money in them, at present!

MATING.

"The sire is half the herd; if he is a poor one he is all of it." Get a vigorous sire. Do not think too much of size.

Look that he is active, muscular, alive all over. See that he is big through the heart. See that he has a straight back, a well sprung rib, a good, short, straight leg. See that his horn is strong, well turned. See that his neck is thick and muscular. Have him well wooled all over. Study your scale of points. Don't quibble about the price, but be a stickler for quality. If he is not right you will regret it all your life, maybe, for it takes ten years' weeding to undo one year's bad breeding. And every year send to the butcher the ewes that you know are inferior.

SCALE OF POINTS.

Adopted by Continental Dorset Club.

HEAD—neat, face white, nostrils large, well covered
on crown and under jaws with wool............ 5

HORNS—small and gracefully curving forward rather
close to jaw 5

EYES—prominent and bright 2

EARS—medium size, covered with short white hair.... 2

NECK—short, symmetrical, strongly set on shoulders,
gradually tapering to junction of head.......... 5

SHOULDERS—broad and full, joining neck forward
and chin backward, with no depression at either
point (important) 15

BRISKET—wide and full, forward, chest full and deep 8

FORE FLANK—quite full, showing little depression
behind shoulder 8

BACK AND LOIN—wide and straight, from which
ribs should spring with a fine, circular arch....... 10

QUARTERS—wide and full, with mutton extending
down to hocks.......................... 10

BELLY—straight on under line.................. 3

FLEECE—medium grade, of even quality, presenting
a smooth surface and extending over belly and
well down on legs 12

GENERAL CONFORMATION—of the mutton type,
body moderately long; short, stout legs, placed
squarely under body, skin pink, appearance attractive 15

Total.............................. 100

DISEASES OF SHEEP.

THIS is not meant for a scientific discussion of diseases and remedies; it is merely an effort to group the common ailments under their common names in alphabetical order. The remedies given are tried ones and the directions brief and simple, just as if we were talking to each other. Where there is no practical, tried remedy known, no attempt is made to appear erudite by naming possible ones. In such cases prevention is indicated, which after all is the Great Sheep Remedy.

A B C Ailments and Remedies.

Abortion—Strictly speaking this is not a disease, but the result of disease or accident. Among sheep it is seldom epidemic. If the contagious form appears the cause must be ascertained in order to apply any remedy for checking the spread. In cases of epidemic abortion you should get the advice of a veterinary. Individual cases are mostly the result of crowding the animals through narrow spaces; rough handling; fright or injury from any cause.

Braxy—This is a disease of sheep; but the term is so variously applied in different sections that it is not wise to specify causes or remedies.

Bloating—Give tablespoonful baking soda, and half tablespoonful ground ginger in pint of water. Fasten mouth open with band of straw or piece of corn cob; straddle the animal and gently but regularly knead or work the extended sides. Use the trocar as a last resort.

Barren Ewes—Too much flesh is usually the cause. Force constant exercise; reduce feed for month or more before breeding; then feed liberally.

Choking—Give small doses of linseed oil, and work throat gently with hand. As a last resort, bunch securely a rag on end of piece of whalebone or other flexible substance, oil, and push carefully down throat.

Casting the Withers—See Prolapsus Uterus.

HOW THEY RUSTLE, SNOW OR NO SNOW.

Diarrhea-Scours—Give teaspoonful to tablespoonful castor oil to suckling lambs. For older sheep change pasture or feed. Protracted cases can be helped by giving handful wheat flour in feed. Diarrhea of weaned lambs and mature sheep is often caused by worms; also by certain weeds in pasture or hay. In such cases the cause must be removed.

Docking—Use the docking irons or pinchers. About a week old is good age to operate. But if the lambs are strong any age from few days to fortnight will do.

Foot Rot—Cut carefully away every particle of the diseased part, and apply salve made of blue stone·and lard. Tie coarse bagging around foot to keep salve on and dirt out. When the rot extends into the flesh above the hoof, wash with a 50 to 1 solution of carbolic acid, and apply powdered burnt alum. Running sheep through fresh slaked lime watered to the consistency of paste is great preventive of foot rot. Dry lime put around feeding and watering places is helpful also.

Garget—This often follows neglected caked-bag, particularly apt to if ewe is exposed to wet and cold. If udders are properly looked after at lambing and weaning time it will seldom occur. For general treatment see Caked-Bag.

Goitre—There is no known remedy that will prevent this serious trouble. Fortunately it seems to be prevalent only in certain parts of this country. As it is hereditary, affected animals should never be used for breeding purposes. Iodine will reduce the swellings.

Constipation—For sucking lambs give castor oil, teaspoonful to tablespoonful, according to age and severity of trouble. For mature sheep use epsom salts, 4 to 6 ounces in pint of warm water. Never give salts when there is evidence of pain. Substitute raw linseed oil, or, better yet, castor oil. The use of stimulants in small quantities, such as brandy, gin or whisky, will increase the action of the cathartics.

Castration—From two to four weeks old is convenient and safe age to operate. Apply an antiseptic after operation, such as the carbolized non-poisonous sheep dips.

Colds—Give teaspoonful carbonate iron, as much quinine as nickel will nicely take up, and wine-glassful of whisky. Repeat every other day for a week or so.

Caked-Bag—Keep udder milked out, and do not allow ewe to

be exposed to cold and wet. Apply to udder a liniment made by mixing 1 quart tincture of arnica, 6 ounces tincture of belladonna, and 4 ounces spirits of camphor. Rub on vigorously with palms of hands. If a ewe has a very large, extended, hard-feeling udder before lambing, do not hesitate to milk it out some. Never change suddenly from low feeding to high feed-

CROSS-BRED DORSET-SHROPSHIRE LAMBS.
Ready for Market on Woodland Farm.

ing in a Dorset ewe; the result is apt to be caked-bag. Too much corn feeding is inducive to this trouble.

Grub in the Head—This term is usually applied to a grub that is laid alive in the sheep's nostrils by the Sheep Gad Fly, during the hot months of summer. The grub works its way upwards, causing much distress and "snotty noses." There is no complete cure. The preventives are any contrivances that help the animal to escape the fly. Tar on the nose repels the fly; but it is a difficult thing to keep the noses always tarred. Shady places and strips of plowed ground to lie on, and long grass are helps to the sheep. Death seldom results, but great distress and worry does. These grubs cannot possibly reach the head proper or brain cell. There is a grub, though, that gets there by way of the spinal canal. There is no practical cure for this kind of grub in the head. It is sure death to the animal. A skillful surgeon might resort to trepanning, but, aside from the expense, this would be a doubtful operation, as often the grubs are three, four or more in number, and lodged in different parts of the brain cell, so all could not be removed.

Gid or Staggers—Some authorities call grub in the head by this name, as in its advanced stages the sheep's brain is affected and it staggers about. These symptoms follow the grub in the head proper, not the grub in the nostrils, which commonly goes by the name of grub in the head. There are several ailments which cause sheep to stagger and stumble, and each of them is often called by this name. Highly fed sheep will sometimes accumulate blood too fast, and it will go to the head, causing this staggering symptom. Generally when the sheep acts this way it is surely going to die, no matter what the cause.

Hoose or Husk—Another name for lung disease. See Paper Skin for treatment.

Impaction—Young suckling lambs are subject to this, especially the richly nourished ones. The milk becomes hard or impacted in the intestines. When a fat lamb hangs its ears and mopes around it is very likely impaction. Give tablespoonful castor oil; if this does not move the bowels, give injection of warm soap suds. For mature sheep give wheat bran, with little salt, made into a mash. For severe cases, two or three ounces each of raw linseed oil and molasses will make a strong purgative. Mature sheep are not subject to this ailment if fed at all properly.

Inflammation of the Stomach--This may follow neglected cases of impaction. The sheep's evidence of distress and pain in the stomach will indicate this trouble. Two

ounces castor oil, with half ounce of laudanum, will relieve somewhat the pain. If there is fever, as is probable, five to ten drops of aconite will be of help. Keep all feed away for day or so, and then feed lightly of succulent, laxative food.

Knotty Guts—See Nodular Disease.

Liver Fluke—This is a disease of the liver caused by internal parasites. In its advanced stage it is hopeless to give medicines. The source of infection must be ascertained and destroyed or sheep removed from same. The drinking of stagnant water is common cause.

Lice—Use any of the standard sheep dips as directed. Dalmatian powder applied with powder gun is effective for red-headed louse.

Lung Disease—This is a disease of the lungs caused by internal parasites. Some writers refer to it as paper skin. The animals lack blood, the skin looks white, also the lips and eyeballs under the lids. As with most other internal parasites, there is no known cure for this. For general treatment of anæmic condition see Paper Skin.

Nodular Disease—Knotty guts: This is a disease caused by internal parasites. An examination of the intestines will reveal numerous little tumors or knots growing to same. Many sheep are more or less affected with this, and no apparent harm results. At times and in certain sections it is very destructive. No positive remedy is known. All that can be done is to give extra care, change of pasture, and avoidance of any possible source of infection. This disease is more common and fatal in the South than in the North.

Paper Skin—Properly, this is the name for lung disease. Generally, though, it is applied to sheep in an anæmic condition, and this condition is the result of various diseases. There is a lack of blood in the system, causing the skin to appear white and lifeless like. Carbonate of iron is a blood builder, and a tonic of this with equal parts each ground ginger and gentian is very excellent for the anæmic condition. A tablespoonful of the mixture once a day in feed for a week or so. A sheep in good health has a pink, inviting skin. When the skin gets pale or white it is a sure sign of some ailment. As soon as this symptom appears start at once and give extra care and attention, and feed with above tonic.

Prolapsus Uterus—Falling or protruding of the womb. Many cases called this are merely the inversion of the vagina— literally the turning inside out of the lining membrane. The rectum also sometimes protrudes. Give laxative foods. Thoroughly cleanse the protruding parts, anoint with raw linseed oil with little laudanum in it, knead gently and return. Give internally four ounces of raw

linseed oil with tablespoonful of laudanum. If above is not successful after a few trials, it will be necessary to fix straps or harness so as to hold the protruding parts in for a few days.

Rheumatism—Young lambs are occasionally affected with a stiffness and lameness of joints. It may be lamb founder; but in early spring is apt to be caused by lambs lying on cold, damp ground. Keep yards well bedded with corn stalks or other roughage, so lambs cannot lie on bare ground.

Scours—See Diarrhea.

Scab—Use any of the standard dips as directed.

Sore Mouth—Rub scab off and apply a non-poisonous carbolic sheep dip. Powdered burnt alum is also effective where it can be made to remain on. The seat of this trouble makes it difficult to apply remedies. It will, however, usually disappear of itself.

Sore Eyes—Cleanse with warm water and drop little witch-hazel in and around eyes.

Scum on Eyes—Usually will disappear without treatment. Introduction of any substance to cut the scum is cruel practice of very doubtful necessity. Relief can be given by washing as for sore eyes.

Snotty Nose—A symptom of Grub in the Head or a Bad Cold. For treatment see both these headings.

Swelling Under the Jaw—This is not a disease, but the symptom of several ailments. Often it accompanies an anæmic condition. For general treatment see Paper Skin.

Scanty Urine—Rams and wethers may have trouble in making water. Give from one-half to one ounce sweet spirits of nitre; put pinch of powdered saltpetre in feed for several days. Do not feed mangels.

Stone in the Bladder—This is another ailment of lambs and wethers. Like goitre, it seems to be largely confined to certain sections of country. There is no positive cure for this ailment. Mangels cause it, and aggravate mild cases, so they should not be fed in any great quantity.

Stomach Worms—There are many kinds of stomach worms. Usually, though, the term is applied to the strongylus contortus, among lambs, one of, if not the most fatal of all internal parasites. An effective remedy is ben-zine or gasoline given in sweet milk for three consecutive days. A dose is from teaspoonful to tablespoonful, according to age of lamb. Add to each dose about half glass of sweet milk. Shake well together. Shut lambs up over night so as to give on empty stomach. Have assistant set lamb on rump when you give the medicine, and be sure he holds head in natural position for the posture, otherwise strangulation may result. See chapter on Parasites.

Ticks—Use any of the standard dips as directed.

Tape Worm—At times and in some localities this worm proves very destructive. Ordinarily, though, a few tape worms seem to be a necessary accompaniment of a lamb's growth and do no harm. If numerous they can be expelled with any recognized vermifuge, such as powdered araca nut in one to two dram doses on empty stomach. Follow in from twelve to twenty-four hours with a cathartic. Pumpkins are good, as the seeds act as vermifuge.

LAMB FOUNDER.

The Lambs so Gentle the Girls Pet Them.

There is a peculiar disease of lambs that causes them to become very stiff in their joints. It may attack one joint or all the limbs may be affected. They lie around a great deal and move painfully. They seldom die, but are checked sadly in their development. The cause of this distressing ailment is to be sought in the ewe. She has been unwisely fed. Most probably she has been allowed to gorge herself on grain, or her feed has been changed abruptly from a light ration to a heavy one. This creates indigestion and a peculiar poison in her system that shows itself most in the lamb. There is no cure but time, and an avoidance of the contributing causes. So far as we have observed a high feeding of corn is most apt to cause this disorder.

While the suckling ewe should be well and even highly fed, she should never be changed suddenly from a light ration to a heavy one, nor should she ever have a large allowance of corn.

BROKEN HORNS

may cause the death of the lamb. Sometimes a sort of blood-poison sets in that causes the head to swell so that the eyes are even swelled shut. There is no help for it but time after the infection has occurred, but if at once when the horn is noticed to be broken the stub be smeared with some carbolic sheep dip there will be no infection and no bad results. One should plan his pen so far as possible to be tight and smooth so as not to catch and break the horns, which are very tender at a certain stage of development.

DOCKING TAILS.

There is but one right way to dock tails, that is with the docking pinchers made by F. S. Burch, 178 Michigan Street, Chicago, Ill. These iron

pinchers are heated to redness and the tails severed; no bleeding occurs, and the tails may be made very short. This is best done at about ten days of age. If there are flies a smear of tar or sheep dip will deter them until the wound is healed. It heals very quickly when the pinchers are used. Take a board six inches square, bore an inch hole through the middle of it, thrust the tail through this hole and cut as close as you can. The board holds the tail and prevents scorching the lamb.

CASTRATION.

For the winter market to go from their mothers' sides it does not matter whether the lambs are castrated or not. Some growers always castrate, others never do. It is probable that if the castration is done carefully and soon enough the lamb may fatten faster than if his testicles were left in. The castration of lambs a week old or less is a simple matter; the end of the scrotum is cut off, the testicles drawn out, cord and all, a little lard and turpentine placed in the wound and in a short time the wound is healed.

Later in the season, when ram lambs have been let go and some have turned out badly and are not fit for breeding rams they are hard to castrate without loss, but the docking pinchers may be used again, taking off the entire scrotum, as you would dock the tail. I have never seen ill results follow this operation, and have castrated rams six years old in this manner. Care should be observed to have the pinchers quite hot.

SORE MOUTH.

There is a contagious sore mouth that affects lambs and sometimes sheep. Warty scabs form on the lips and nose, making it difficult for the lambs to eat. Similar sores appear

A Grade " Rent Payer."

on the ewe's teats. The cure is simple. Rub off the scales and apply some carbolic sheep dip. Milk-oil, made by F. S. Burch, 178 Michigan Street, Chicago, has proved effective in the experience of the writer, and one application has always been sufficient. There is another form of this sore mouth that is confined to the lambs. It is more difficult to cure, but after running its course will quickly disappear.

SORE EYES.

The contagious sore eyes that sometimes appear among the flock in Winter are also easily and quickly cured by a tiny drop of sheep dip dropped in the eye. It should be diluted about ten times with water and not only a little allowed to penetrate the eye, but the face should be scrubbed with it, especially wherever the tears have run down the cheeks. There is no excuse for allowing these and similar petty diseases to spread and become formidable; a little watchful care, a little disinfecting with carbolic dip, and the disease is cured and its spread stopped. Milk-oil, or some similar carbolic preparation, should always be at hand in a bottle; or, better, an oil-can on a shelf in the sheep barn.

INTERNAL PARASITES.

Would it were as easy to keep the inside of a sheep clean as the outside. Unfortunately this is not so. Sheep suffer greatly from a number of internal parasites, but in America the chief and almost only important one is the tiny stomach-worm. The lambs that are dropped in fall and kept on clean pastures until cold weather are seldom troubled with these pests; the lambs dropped early in winter and fattened and sold before June are safe, but the late lambs that must run with their mothers on grass are apt to become affected. The symptoms are a general lack of thrift, a sunken condition of the fleece, a paleness of the skin, the eating of earth and rotten wood, a slight cough, sometimes scours, at other times constipation, emaciation and often death.

One should never see a lamb die on his place without dissecting it to learn the cause. If it is stomach-worms, they may be easily found in the small fourth stomach, the place where the intestines begin. Stomach-worms are small, hair-like worms, about three-fourths of an inch long, twisted in the middle, from which they take their name, Strongylus Contortus. They may be present in sheep having apparent good health; they may even in small numbers distress the lambs; they may be found in immense multitudes, blocking the intestinal canal. They seem to greatly disturb the digestion and assimilation, and no lamb will thrive with these pests within him.

The infection is nearly always from the grass or from stagnant water fouled by sheep's excrements. The ewes are apt to be slightly affected, the worms discharge immense numbers of eggs, perhaps at all seasons, certainly in spring and summer. The immature worms in some way cling to the grass and are taken in by the lambs when grazing. In some mysterious way nature aids the older and stronger sheep to throw off most of these pests, while the smaller and weaker lambs become affected very easily. The lesson is that all small, grassy yards, where sheep love to lie and where the grass is thick and tender, are unsafe, almost surely fatal to the lambs. Unfortunately the short, sweet grasses, such as Kentucky blue-grass and white clover, are the very worst and most dangerous from the point of infection, as the sheep bite them so close. Red clover, alfalfa, orchard grass, bromus-inermis, timothy, oats and barley and rape, are all bitten higher up, and there is much less risk of infection. Also in soiling sheep there is hardly any danger if the racks are not soiled by the sheep excrements. On Woodland Farm there has been a notable decrease, almost a disappearance of this pest, since alfalfa pasture has been the main reliance. It is also a good plan to let the sheep shade in the barn, as then their droppings are not soiling the grass about some shady tree, where the grass will grow up rank and sweet and be nibbled at by the unsuspecting lambs with fatal re sults. Care should at all times be taken that the sheep should not drink from stagnant pools or small, slow streams fouled by the droppings. Troughs are much the safest watering places, and they should be kept clean.

As to medication, it should be prompt upon the first sign of infection. The old remedy of turpentine and milk is

Dorset and Shrop Blood Mingled.

rarely effectual. It is not worth administering. The only things that have seemed to do good are Toxaline, a preparation made by F. S. Burch, of Chicago, and gasoline or benzine, which was discovered by M. Julienne, in France, and first introduced by us into America. In case infection is discovered it is wise to treat the entire flock. Be careful not to strangle the sheep by rough or too hasty drenching.

Either benzine or gasoline may be used. The dose is two teaspoonfuls to a 50-lb. lamb, mixed with four ounces of either sweet milk or thin flaxseed tea (cold), well shaken together. Give after fasting for 16 hours. Be careful not to strangle by pouring down too fast or getting in windpipe. Repeat the dose daily for three days. It has no ill effects on the health of well lambs, and is sure to remedy the drooping ones if stomach worms are the cause of their illness. Dose the old sheep as well. They will take a tablespoonful. Better to use a 5c measuring glass (sold at druggists) rather than try to measure in a spoon, which holds an uncertain amount.

GRADE DORSET EWES.

While pure-bred Dorsets are extremely profitable to those who will give them care, and while there must of course be breeders of registered stock to supply the need of Dorset rams, yet it must be remembered that the grade ewe is the rent-payer, the money-maker, and in common hands more profitable than the pure-bred ewe. Indeed, there are some curious things about the grade ewe. If she is a Merino grade, from large, roomy Merino ewes and blocky, vigorous Dorset ram, she will prove a surer breeder, if possible, than the pure-bred Dorset ewe. In truth, not many growers of winter hothouse lambs but prefer Dorset grades from the Merino foundation to any other ewe, the pure-bred ewe not excepted. These ewes are again bred to pure-bred Dorset rams, and the result is a very blocky, easily fattened lamb, born at the right season and ripe for the right market. These grade ewes are great milkers and hardier than pure-bred ewes, and altogether more desirable for mutton-makers. There will come a time when ranchmen will make a specialty of producing ewes of this type, as there is already a demand for them in all the early lamb-producing regions, and they are hard to buy. These grade Dorset ewes will continue profitable for at least ten years and often longer.

Another very profitable grade Dorset ewe is the Dorset-Shropshire grade. This is a magnificent ewe, lambs early, but not quite so regularly as the Dorset-Merino, is a better mother than the Shropshire, with more milk. Ewes of this cross are becoming quite common now. They are usually white or light brown in face and hornless. Sometimes the three-quarter blood Dorset-Shropshires have horns. These are better ewes than the first cross, having, indeed, many of the best characteristics of the pure-bred Dorsets.

USING GRADE RAMS.

However profitable grade ewes may be, it is never safe to use grade rams. They will revert in unaccountable ways to remote ancestors, and there is simply no telling what the product will be.

Of grade Dorsets John B. Peelle, a famous hothouse lamb grower, says:

"The grade Dorset with me is not a question of sentiment, but one of business. The questions I ask of a ewe are: Can you produce lambs in November or December? Can you produce one or more at a time? Can you provide the lambs with an abundance of milk, so that they will be ready for market in sixty or seventy days? It is only the ewe that can answer to all these questions 'yes' that is at all desirable as a mother of hothouse lambs."

John B. Peelle's Man Utilizing the Ewe's Spare Milk After Her Lamb Has Gone to Market.

The first question is most important of all. The best and only remunerative market for hothouse lambs is during the first ten weeks of the year, so the lambs must be here before the snow flies if we want large profits. The October lamb is too early and will only sell as a lamb and not as a fancy product, and the late January and February lambs are too late in the season for the high prices.

So far as I have been able to learn the Dorsets and their grades are the only breed of sheep that will breed with any reliability at the right time. The hot weather that causes most sheep to miss the oestrual period does not seem to affect the Dorsets. In fact, the mating often occurs during the hottest of hot weather. Twenty-four lambs is the record of one of our Dorset rams on one of the hottest of June days, and this occurred in a flock of fifty ewes.

It is not claimed that no other variety of sheep will breed in hot weather, but that the Dorsets will breed more readily and uniformly than any other. If the lambs come scattering along all winter they are a constant care and worry, but when they come in a shower, as ours usually do, it is a pleasure to care for them.

In regard to the number of lambs produced, single lambs from mature ewes are the exception. Triplets are common. Thomas Shaw says: "The Dorsets will probably drop and raise more lambs than any other breed."

As Milkers—The Dorset or grade always has an abundance of milk. Some are such persistent milkers that it takes several weeks to dry them up, but this is a good thing for pets and thieves (see cut preceding page). It is easy to teach the lambs whose supply is short to come at the call. I mean the twins, triplets, and those whose mothers are out of condition, and then, while you hold the ewe, the lambs do the rest. Often the best milking ewe can be made to raise another lamb after her own goes to market.

Recapitulation—The virtues of the grade Dorset may be summed up as follows: She has size and that counts when she is put on the market as mutton. She has constitution and vigor, and that means long life and lots of service. A nine-year-old gummer raised the best part of lambs we had this year. It is no burden to shell corn for her. She is a good rustler. One season's experience showed us that the Dorsets and Merinos have no business in the same barn.

The Merinos simply had no show in the rush for feed. She will produce her lambs at the proper time for them to reach the market when prices are highest. She is a fluent milker, the more milk the quicker the lamb goes to market. Our best ewes, if perchance they have single lambs, will have them ready for market in less than fifty days.

BUILDINGS.

One can do with a very common and cheap building or he can use a good, warm, convenient building to good advantage. The more expensive buildings are needed in the cold, frozen North; in the South very slight protection against cold is needed, but wet is to be guarded against. A safe rule is NEVER to allow the flock to suffer a wetting, unless the wool is very short at the time. It is a very depressing thing to a sheep to carry about a wet fleece, and unfortunately not all or many sheep know enough to come in out of the rain. The illustration of the barn at Woodland Farm (page 12) shows one type well adapted to a Southern situation. The barn is 36 feet square, 18 feet to the eaves, with a half-pitch roof and an open center.

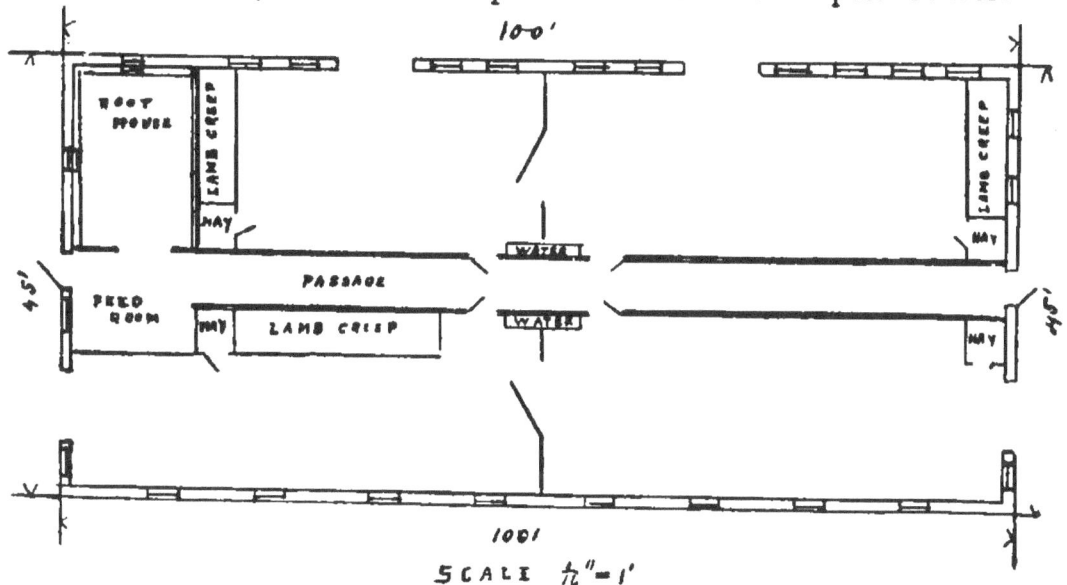

Plan of Barn at Fillmore Farms.

The lower story is 8 feet in the clear and divided by
means of racks into compartments as desired. It will be
seen that it is light and airy and cool in summer, and when
the doors are let down it is fairly warm in winter. This
barn cost to build less than $200, with a good shingle
roof, no floor but natural earth below and rough flooring
for the mow. It accommodates fairly well about 75 ewes
and their lambs. The hay is taken in from the end and
the open doorway is turned to the southeast, so that little
or no storm ever blows in. It could easily be closed,
however.

THE BARN AT FILMORE FARMS.

Fillmore Farms (W. G. Appleby, Manager, Benning-
ton, Vt.), Mr. Colegate's place, has an ideal large barn for
cool climate. The ground plan shows quite clearly the
arrangement of the lower story, 45x100 feet. This barn
shuts up tight in cold weather, four ventilation shafts run
up the purlin posts and then to cupolas, taking off the foul

air and not making drafts. The outer doors slide, and there are slatted doors that also slide up out of the way; when it is warm the solid doors are back and the slatted ones in place. The root house is not a cellar, though dug down the depth of the foundation, but is double boarded, with paper between and two air spaces, and is frost proof. It is convenient to store wool in, and in the feed room is a good shearing floor. The passage is a handy place to pen and catch sheep when shearing.

The feed racks used on Fillmore Farms, the Tranquillity

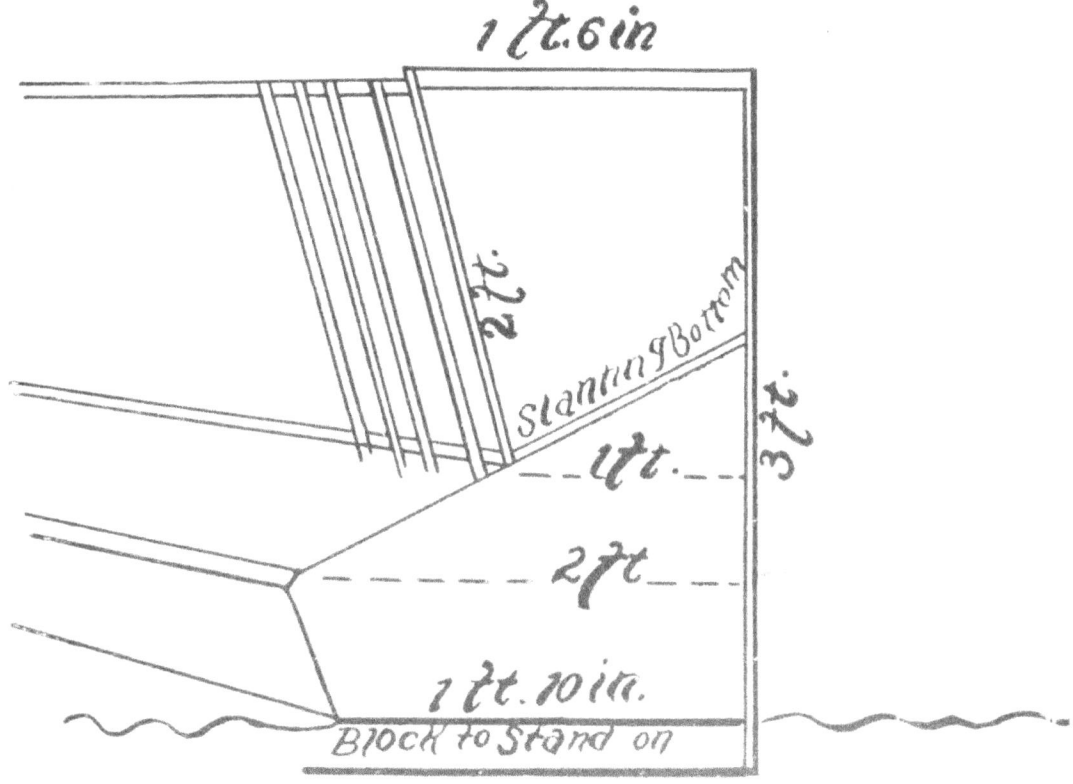

Diagram of Combination Feeder.

Farms and some other places are illustrated herewith. There is probably nothing better for Dorsets, as the lambs cannot soil the hay nor are horns broken in this rack. Here are the specifications:

Trough is 6 inches wide at bottom—14 inches at top, on slant.

Trough is 7 inches high at front—11 inches at back.

Slats 2 inches wide, 1 inch thick, rounded slightly at corners.

Spaces between slats 3 inches.

Slanting board at bottom of rack 1 inch thick.

Slats are of hard wood; rest of trough may be soft wood or not, according to price, etc.

Front board of trough is beveled at top.

Frame 2x3 or 3x3.

Trough may be made any length to fit spaces, or in 8, 10 or 12 foot lengths, to be easily moved around, and back to back they make partition with feed trough and rack on each side; or can be put out doors and make yard with rack, etc.

Cheapest in end. Last forever. No waste hay. Feed roots, grain or anything without loss.

Combination Feeder.

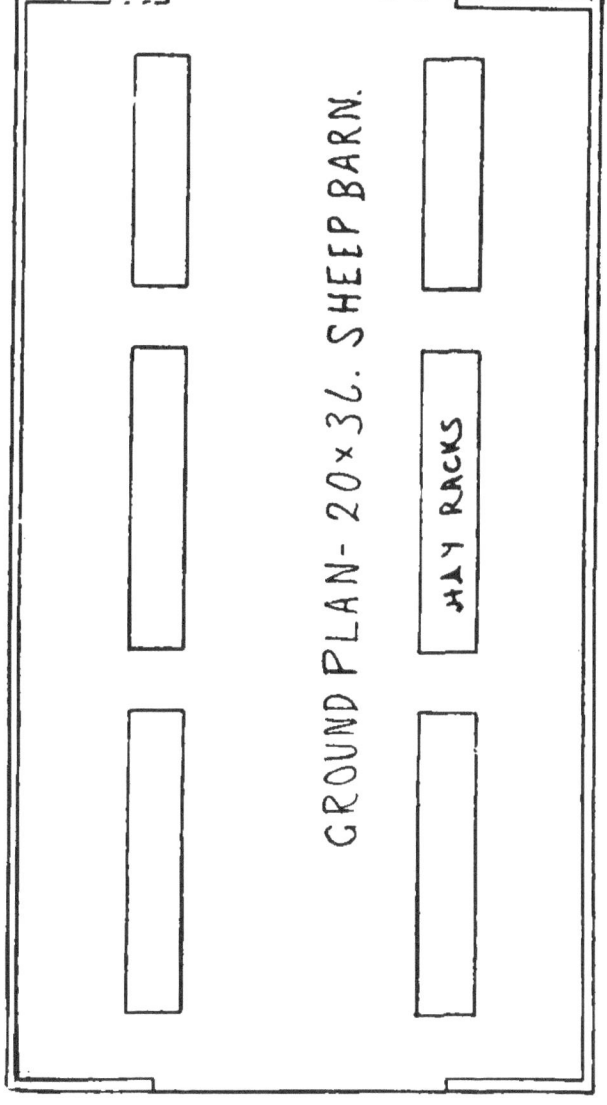

Floor Plan of Ideal Sheep House.

The Ideal Low-Cost Sheep House.

For the following plan we are indebted to the Breeder's Gazette, where it is illustrated in issue of April 10, 1901.

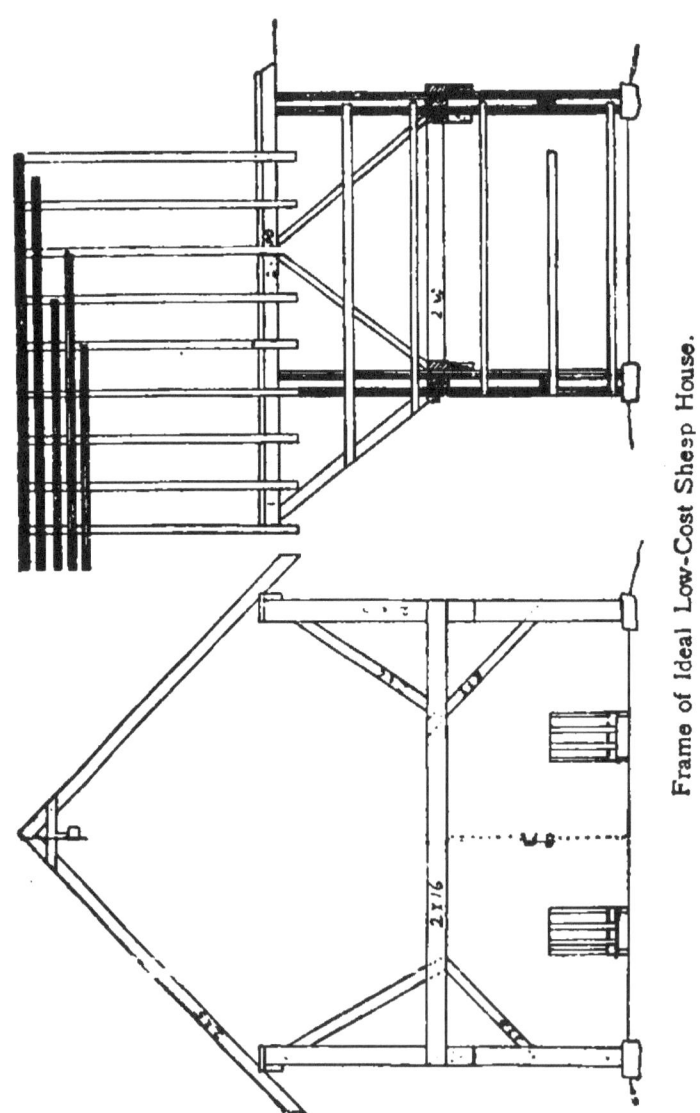

Frame of Ideal Low-Cost Sheep House.

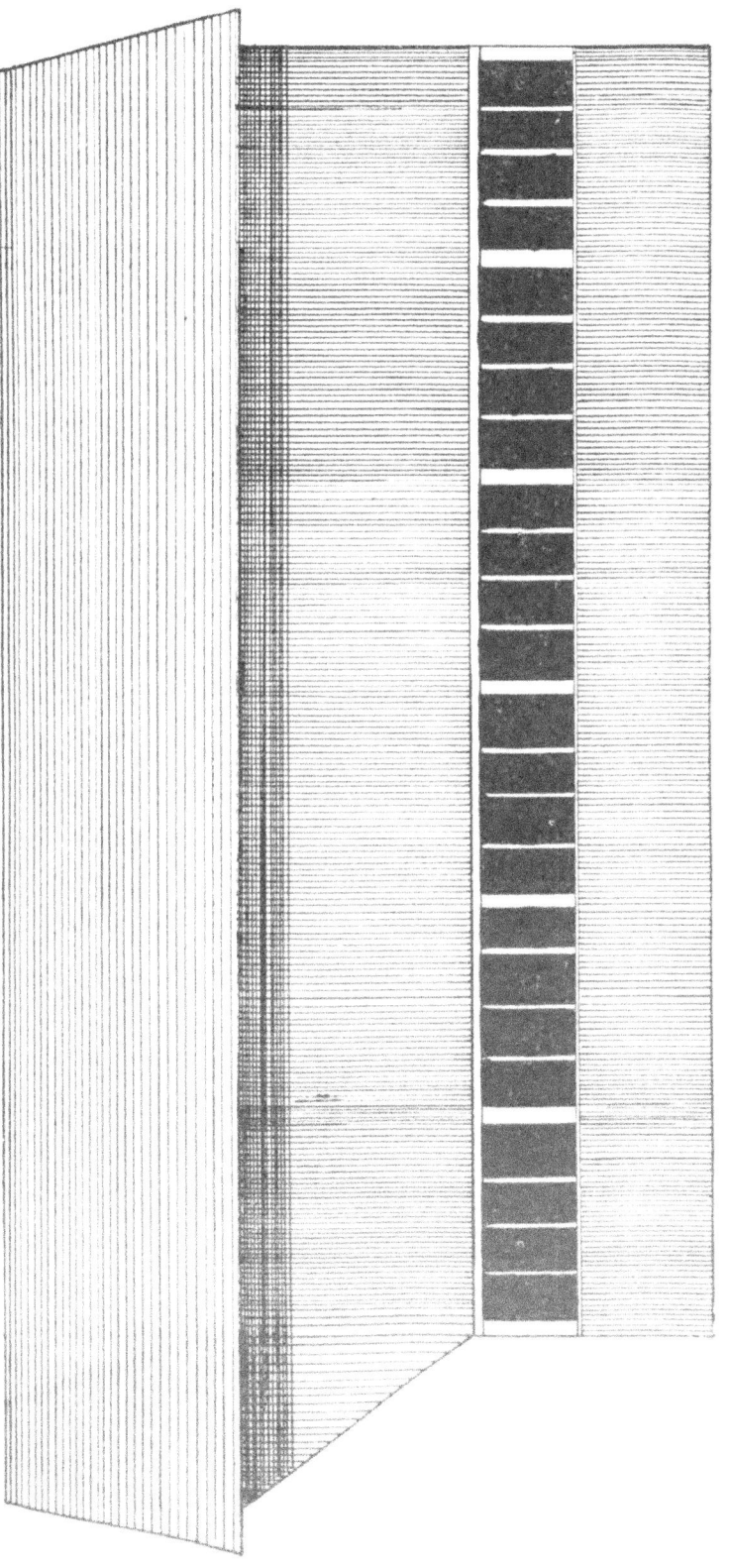

Ideal Low-Cost Sheep House.

It combines low cost with convenience and usefulness of high order. The building is 20 feet wide, as long as desired, 16 feet high at the eaves, with a lower story 8 feet in the clear and an upper story with half-pitch roof and 15 feet in height at the peak. There are no obstructing cross-ties and the hay carrier works on the track in the peak without hindrance. The floor joists are put in lengthways of the building and are of 2x8 or 2x10 while the joist-bearers are 2x16, and the manner of spiking through the joist-bearers into the ends of the floor joists making the upper edges flush saves quite a good deal of head room. This is clearly illustrated in the cut; the floor joists are spiked to the joist-bearers before it is let down to place, then all is firmly spiked together. No floor is used but the hard earth, which is better if rounded up a little so that water will run away from all sides, and a generous projection of rafters helps the appearance and the usefulness materially. There are no divisions to the house except such as are made by the placing of racks or panels across. Hay is thrown down at convenient places through chutes reaching up to the rafters, and at the bottom a pen of hurdles should restrain the sheep from getting on the hay as it is thrown down.

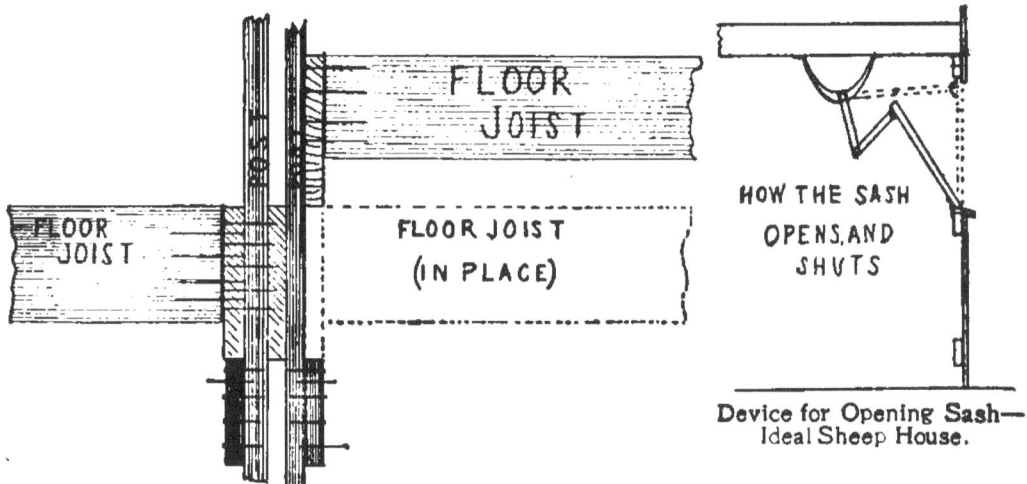

Putting in Floor Joists—Ideal Sheep House.

Device for Opening Sash—
Ideal Sheep House.

At each side there is a continuous window three feet high with sash and glass hinged at the bottom and opening inwardly so as to allow a continuous stream of air to pass over the sheep. These sash are fastened by means of a jointed rod to a continuous rod running through the barn, the familiar green-house sash fixtures, and by turning the rod all the sash are opened at once, either a tiny crack or wide to let the summer breezes through. While this feature may be omitted and wooden windows opening at the bottom and hinged at the top be substituted, yet I think the satisfaction of having it right will in a short time pay for the sash and fixtures, which may be had of any builder of green-houses.

Doors permit the driving of teams clear through the building to clean out the manure and the over-hang of the roof protects the upper doorway through which hay is taken. This building finished well, 20x60, should cost about $300.

Dorsets for Crossing and on the Range.

While the Down ram on the Dorset ewe gets fine lambs, yet the reverse cross is as good, Dorset sire on Down ewe. The lambs of this cross grow very rapidly and fatten very easily and are usually hornless with white or gray faces. A neighbor who used one of our Dorset rams on pure-bred Shropshire ewes lambed in May, sold the lambs before Christmas at 115 pounds average weight, from grass with a trifle of grain at the latter end of their feeding.

On the range the Dorset is new yet, but wherever tried the lambs, usually from grade Merino ewes, have been more than satisfactory. In Colorado, L. E. Thompson, of Las Animas, reports that his grade Dorset lambs are first to become fat and go to market. They are in demand among Colorado lamb feeders wherever they have become

known.   The Range Valley Cattle Co., of Woodside, Utah,
has this season a lot of grade Dorset lambs, in comparison
with lambs from Shropshire and Rambouillet sires.   The
Dorsets are much the most blocky, smooth and heaviest, the
best sellers.   There is need, however, of care in taking
Dorset rams to the ranges; they ought to be young and they

Prize Winners at the Royal.   Dorset-Shropshire Cross Breds.

ought to be sent to the "buck herd" at least a month before
needed to get accustomed to range life.   It would even be
better if they were grown on the range.   There is no sheep
such a rustler as the Dorset, and this makes him pre-
eminently suited to range conditions.   The ranch that is
marketing feeders or fat lambs cannot afford to overlook
the Dorset.

There is another use for the Dorset blood on the range to
infuse good milking properties into Merino flocks.   Even
a small percent of Dorset blood in the ewe makes her a
good mother and profuse milker, and her lamb therefore a
good one.   It is unfortunately true that many flocks of
Merinos are poor milkers, some so markedly so as to be
quite unprofitable; there an infusion of Dorset blood

will work a revolution, for while it will not cut down the pounds of scoured wool it will add to the size, hardiness and mutton qualities of the flock. The grade Dorset ewe has proved herself in Texas, Colorado, Utah, Wyoming, California, Washington and Oregon to be the best of all rustlers and mothers, producing always the strongest and largest lambs.

AS THE EWE LAMBS

The Ewe that Won Over All Breeds at Omaha.

it is well to have some close pens for use in cold weather; these may be made of matched lumber 6 or 8 feet square, and a very little ventilation will suffice. By putting a ewe in here and hanging up a lantern above her, her lamb will not chill in the coldest night. Sometimes one can tell when a ewe is about to lamb and separate her from the flock. This is not always practicable, however, for many ewes will apparently be all ready for weeks and others that have made little show will lamb before them. Let them alone at lambing time, yet keep near by and watch them. If they have not been frightened by dogs or crowded through gates they may not have wrong presentations and the lambing gives no trouble, especially is the danger of trouble lessened if the ewes are strong but not too fat and have plenty of out-door exercise. Yet now and then a lamb will be presented wrong and your help will be needed. The right presentation is head first, front feet on each side. Hind feet first can be taken with no harm to lamb or ewe. Other presentations must be straightened out. Don't wait too long to do this; be slow, careful, think what you are doing, use plenty of lard on your hand, and you may save a valuable ewe and her lamb, too.

A CHILLED LAMB

will be found now and then. If it is too stiff to suck take it at once to a large bucket of hot water, not warm water, but hot as you can bear your hand in. Immerse him all but his nose. Put in more hot water after he has cooled it off. I have revived them after they were apparently dead. Dry thoroughly, perhaps give a wee drop of whisky, then the mother's milk. Don't ever give anything but the mother's milk if you can avoid it. If you must give cow's milk, dilute it half with warm water and add a tiny bit of sugar. A lamb that can't have a good lot of some ewe's first milk is not apt to live.

Another handy thing for lambs not so badly chilled is a half barrel with a lantern or jug of hot water in it and a blanket thrown across it. When the ewe has twins you can keep one warm while she is licking off its mate. Once dry and full of milk there is not much danger of chilling in the most severe weather.

In lambing a lot of ewes in cold weather one loses very few lambs if he will go to the fold at 10 in the evening and again at 4 in the morning. If the lambs are sired by a vigorous ram, and the ewes treated right, not one lamb in ten will need your attention or help. First lambs are often more trouble. But remember, ALWAYS milk out the ewe after the lamb has filled up, and do this not once but daily for a week if she has surplus milk. Stagnant milk in a gorged udder is surely fatal to the lamb. And Dorset ewes, if rightly fed, are great milkers. It is more trouble at first but when the lamb does take it, what a pleasure to see him swell and grow!

If you have a large number of ewes to lamb in winter you should provide a lot of pens, about 4 or 6 feet square. These are best made of little panels 3 feet high and 4 feet long, hinge two of them together at one end and then they shut up and lay away until needed, when they are opened out and hooked to the corner of the barn, enclosing a space 4 or 6 feet square. Another pen goes

Panels Opened   To Make Pen.

alongside, and so on as there is need. Ewes with twin lambs ought always to have one of these pens to keep her family together until they know her.

TRANSFERRING LAMBS.

Supposing you have a ewe that loses her young lamb, you should at once remove its skin, taking it off as near whole as you can, rub it dry on the flesh side and sprinkle it with salt, take a twin lamb that needs more milk and slip it into this skin, put the ewe and odd lamb in a pen together, and the chances are mighty good that she will adopt the stranger with joy. After a few days the skin may be removed, though it is well to take it off a piece at a time.

If the ewe has a large lamb to die this plan may not work, but to put her in a pen and confine her head between stanchions, which may be two small round stakes driven into the earth and confined at the top with a cord, will be the surest and easiest plan. Turn the lamb with her; she cannot refuse to let it suck. After a time, when her milk has given it a new odor, she will own it. This takes from two days to a week.

ALWAYS separate the ewes with lambs from the ewes yet to lamb. You can't feed the same ration to each lot with success.

Woodland Dorsets on Alfalfa.

THE WINTER LAMB.

By H. P. Miller.

About ten years ago I first learned that a few men in New York were raising what were called "hot-house" lambs, which they sold at what seemed to me fabulous prices. They were said to get eight to ten dollars each for lambs as many weeks old when hog-dressed and sent to New York City. I thought the demand only a passing one and that the supply would soon exceed it, so was slow to engage in it. There were other reasons for my delay.

Dressing them seemed to demand the services of an expert. I could dress a sheep for our own use, but I was not an expert butcher. Then the distance from Central Ohio to New York City seemed too great to safely send dressed meat. Again I did not see how I could find a market for them. I did not realize that they were a regular product upon the provision market and could be sold through commission ·merchants. But the prices that were obtained year after year by those engaged in growing this product incited me to read everything I could find about the business. I found that the market was increasing, that I was only a few hours from New York, that some commission merchants were as honorable as men in other business. I finally had the whole theory but did not have the lambs. We had from my earliest recollection been growing Merinos of the Delaine type, having the lambs born in March and April. We let a flock of ewes run over one fall without breeding, and turned the ram with them the following Spring. We found they conceived as well in May as in October. Bear in mind our ewes at that time were all pure Merinos of the Delaine type. A further surprise, and one quite as agreeable, was that the lambs born in the fall grew more rapidly and when sent to market at three to four months old brought more than lambs of the same breeding born six to eight months earlier. Still we did not get the prices we had read about. We used a mutton sire of a Down breed, but our lambs were not prime. The lambs had too much wool before they had size or were fat enough. The ewes were not good enough sucklers to make the lambs choice. It seemed reasonable that the pure or grade mutton breeds would be more profitable for this business. We provided ourselves with small flocks of three of the leading mutton breeds, but February, with an occasional lamb in January, was as early as we could get lambs from them. That was not early enough for best prices.

One year I tried twenty-five young Merino-Southdown ewes, putting them with an equal number of pure Merinos, and turned ram with them in June. The first lamb from the cross-bred ewes was dropped in March, after most of the lambs from the Merinos were marketed.

We had before this learned the merits of the Dorset and had secured a ram. The half-blood lambs pleased us in appearance and in profits. The next step was to get some half-blood ewes. We have them, use them and are satisfied. The Merino-Dorset ewe is the right one for growing winter lambs.

May is a favorite month with me for breeding. I would prefer to breed a month or six weeks later, but the ewes or ram, or both, are not so favorable to it. I would prefer not to have the lambs born until the ewes go to their winter quarters. The lambs then entirely escape the stomach worms and they can be gotten to eating grain younger. The only special treatment I have found necessary to induce the ewes to breed is such care as will insure improvement in condition. They do not need to be fat, but should be GETTING fat. Indeed, I have found it advisable to put the ewes on a very light, dry ration as their lambs are slaughtered so as to reduce their condition. Then remove their fleeces with the first settled warm weather in April and turn upon good pasture.

I endorse the recommendation given on page 8 in reference to breeding, but it is not always practical to remove the ram every morning and return him to the flock in the evening. You can change rams once a week, or if rams are cheaper than your own time, place two with the flock at once. Jealousy will incite them to watch the flock closely.

This, of course, is hard on the rams, and recommended only as an expedient. At this point re-read the chapter on "Summer Care of Pregnant Ewes." As the lambs appear remove them with their mothers from the main flock. With the Dorset and grade Dorset ewes, if they should have but a single lamb their udders will need to be watched for the first week and surplus milk removed. There will nearly always be some lambs in the flock that will need it and will quickly learn to take it as shown on page 49. As soon as the lambs are taking all of their mother's milk feed the ewes to produce all the milk they will take. The ewes need a milk cow's ration. So long as the grass remains good, supplement it with corn, oats or barley and wheat bran; or substitute for the latter the gluten feed in small quantity. It is worth about twice what wheat bran is to feed in connection with corn and should be mixed with corn in proportion of one to two.

The lambs will begin to eat at about three weeks of age, some of them younger. There is nothing they like better than cracked corn and wheat bran. We occasionally add to this combination oats, barley or gluten meal or feed. A variety induces them to eat more, and the more the better at this early age. I have never known one to over eat. We formerly used a self-feeder—that is, a trough so devised that the feed becomes accessible as fast as eaten, but have discarded it as the feed is liable to become foul. We find the lambs do better if the feed is given them fresh in a clean trough three times a day. Nice clover hay is almost indispensable for both ewes and lambs. Alfalfa or soy bean hay may be substituted. The lambs must not be compelled to eat their hay close. It must be changed three or four times a day. The little lambs as well as their mothers need to have both salt and water accessible. We have fed ensilage three winters with entire satisfaction. Indeed, we would not think of trying to raise winter lambs without it.

It is altogether the cheapest feed we can prepare, is relished perfectly by the sheep and little lambs as well, and it makes fat lambs. We feed it twice a day with a little gluten meal sprinkled over it. With a light feed of hay once a day this constitutes the entire ration of the ewes after they go to winter quarters. Neither the ewes nor the lambs leave the barn from the time they enter it until after the lambs are slaughtered. Some other nitrogenous feed might be substituted for the gluten meal. The determining consideration is the cost. The term "hot-house" formerly applied to these early lambs led many people to think they must have an artificially heated house. This is not necessary. They need a stable into which the wind cannot blow, one with considerable glass on the south and west sides. But the stable does not want to be closed all the time, only, indeed, upon very cold days. Pure air is essential. If the air is ever noticeably foul on entering the stable from the outside get some pure air into it at once. To prevent the escape of ammonia from the accumulating manure there is nothing else as effective as acidulated phosphate rock, just such as is sold for fertilizer. The free sulphuric acid in it combines with the ammonia in a somewhat stable combination, yet one that is available as plant food, so that the fertilizer is not lost. Bedding should be used in sufficient quantity only to keep the stable clean. Any excess encourages heating. If possible have the lambs' private apartment where they are fed hay and grain so situated that the direct rays of the midday sun fall into it. This should be shut off from the old sheep by a creep. Make this of slats placed perpendicularly 8 to 10 inches apart. Let nothing disturb the lambs or their mothers. The lambs should do nothing but eat and sleep, not even play. During the early part of the season forty-five pounds live weight is large enough. But weight is not the only consideration. They must be fat. There is a very limited call for them for Christmas dinner, but the general demand opens after the people have recovered from the poultry glut of the holidays. The demand for them continues strong until settled warm weather.

Arrangements should be made with some reliable commission firm unless fortunate enough to get a good private customer. The commission charged is 5 per cent. They

A Pen of Royal Winners.—(Courtesy Farmers' Advocate.)

must be shipped to arrive at the commission store as early as Friday morning. We formerly shipped by express at the rate of $1.50 per hundred, but the past year they went through in equally good condition by refrigerator freight at just half the charge.

The preparation for market requires some skill, yet only such as almost anyone can develop after carefully studying directions. We have greatly simplified our method of preparation and the lambs apparently sell as well.

It is very important to have them thoroughly bled out. To secure this we have found it advantageous to suspend the lamb by the hind feet in killing. Suspend a short single-tree about six feet from the ground. Loop a small rope or strong twine about each hind leg and attach to the hooks of the single-tree. With a sharp-pointed knife sever the artery and vein in the neck close above the head. Be sure to sever the artery. Bright red blood is the assurance. The venous blood is dark. Severing the head with a broad-ax would perhaps cause less suffering and insure a thorough bleeding. I remove the head with a knife as soon as the lamb quits struggling. Clip the wool from the brisket and strip four or five inches wide upward to the udder or scrotum, also from between the hind legs as in tagging sheep. Now open the lamb from the tail to the brisket. Slit the skin up the inside of the hindquarters about four inches and loosen the skin from the underlying muscles for two inches on either side of the openings in the skin for the attachment of the caul fat. This should be removed from the stomachs before they are detached, and in very cold weather placed in warm water until ready to be used. Next remove the stomach and intestines. In the early part of the season the liver, heart and lungs are not removed, but when the weather gets warm they must be. Carefully spread the caul fat over all the exposed flesh. Good toothpicks should be provided for attaching it and holding in place. Make small slits in it over the kidneys and pull them through. This part of the work is the one that requires skill to make the carcass look attractive. Now hang it in a cool place for 12 to 24 hours. In extremely cold weather 12 hours will be enough, but better make 24 the rule. Then neatly sew a square yard of clean muslin about each lamb so as to cover all exposed surface. We formerly wrapped each one in burlap and attached to a stretcher, but now place three in a light crate and tack the burlap over the top. We line the crate with heavy paper. Prepare them as shortly before shipping as possible. In warm weather ice may be put between the lambs, not in them. Send them as they are ready, three or six at a time. The market varies greatly, depending upon weather and the number arriving. It is useless to try to get them all in on a high market. Aim to slaughter regularly each week if you have lambs in condition, and keep your commission firm informed as to how many you will probably send and when.

Attention to details is the secret of success.

A SYSTEM OF MANAGEMENT THAT INSURES A HEALTHY FLOCK

Two men in America fought stomach worms all through the disastrous years of the 90s, when little was known to help; they found light, they conquered the pests in a measure, and kept on keeping sheep and studying flock management. Finally each made a journey to England and studied the conditions there with a view to solving the problem for America. There they found hurdling the best answer to the question. Independently of each other they reached the same conclusions as to the practical solution of the question in America. Dr. H. B. Arbuckle of West Virginia and the writer were the two men. But they wish to give all due credit to the Department of Zoology of the Bureau of Animal Industry at Washington for at least giving accurate details of the history of the Hæmonchus contortus (formerely called Strongylus contortus) for without the details that we now have no certain plan could have been formulated.

The basis of this plan is the fact that lambs are born free from parasitic infection; they are healthy. It is only necessary to keep them healthy by preventing infection. Their mothers carry over in their bodies the germs that will infect them in the form of mature stomach worms, which when ripe pass away in the droppings and thus infect the pasture. When the temperature is below 40 degrees the eggs will not hatch. When it is above that they will hatch out in a few hours or in a week or so, depending upon how warm it is. Freezing or drying soon kills the unhatched eggs. So it is seen that ewes will not pollute a field in winter, their droppings are sure to be soon frozen, at least in the region where sheep are mostly kept. But if the tiny worm hatches from the egg it feeds for a time upon the material of the manure and continues to grow till it is about one-thirtieth of an inch long. Then it creeps up on a blade of grass and waits to be swallowed by some lamb, after that it finishes its growth within the fourth stomach of the lamb, and, incidentally, finishes the lamb as well.

Now how to manage a flock with safety and profit on natural grass. To begin with the ewe flock should be treated for stomach worms. This is best done in the fall, when they come from pasture. It may be again done in the spring before their lambs come. Remedies for treatment will be found under the heading "Diseases of Sheep." The writer is of the opinion that use of some of the coal tar dips, in small doses, much diluted, will eventually be recognized as most efficient. This treatment alone has doubled the weight of lambs in some experiments in Kentucky. Next, the flock should at the approach of spring weather be confined to the yard and shed. There are two reasons for this; the one that it is better for the grass, and thus in the long run better for the flock, and the other that there is no contamination of land over which the lambs will later feed. If it were possible

to wholly eradicate the worms from the ewes by treatment
this care would not be needed, but unfortunately it seems
almost impossible with our present knowledge to kill all of
the worms by any medication. While confined to the yard
the lambs will probably be born. It is essential that the
flock be well fed at this time so that the ewes will be full
of milk. If desired there may be provided a run to a rye
field, or to some grass pasture that will not be afterwards
used that summer, to help stimulate the milk flow. By
May 15 probably the grass will be so forward that the flock
may be turned out for good. Now begins the new management. Instead of turning the flock to a large pasture to
roam over it at will turn them on a very small part of it.
How best to manage this will depend upon circumstances.
The writer thinks that in our land of small supply of labor
and much hurry and turmoil during the summer season it
is safest to divide the pastures by permanent wire fences.
These are not costly and need not be very high. We will,
then, turn the whole flock together into the first division;
none shall be scattered about. Of course there may be two
flocks, one with lambs and a dry flock, but the dry flock
had better be put apart somewhere or else put with the
ewes. It will not do to let anything interfere with the regular rotation of these pastures. Now once in this pasture
the flock will be allowed to eat it down close to the ground.
That will not hurt the grass, for all will go in a short time
and the grass may spring up again. This is how pastures
are often managed in England by hurdles.

Dr. Ransom says that sheep may probably be safely left
on May pasture for two weeks. We will shorten this time
to ten days, to make sure. That is, the germs falling to
the earth could not before ten days find their way back into
any sheep or lamb, and we are going to move the flock on
before they are able to get in.

Now in the division between this pasture and the next we will place creeps so fixed that the lambs can readily pass through to the next enclosure. This they will early learn to do, and so they will be eating the fresher parts of the herbage in advance of the ewes.

In ten days then the whole flock will go forward one pasture, the lambs yet having access to the fresher feeding on ahead. Doctor Ransom says we will need for this sure treatment the following divisions:

For May, 2 pastures.

For June, 4 pastures.

For July, 4 pastures.

For August, 4 pastures.

For September, 3 pastures.

For October, 2 pastures.

That makes nineteen enclosures in all and insures that the flock shall be kept in absolute freedom from infections throughout the year.

However, one will not absolutely need so many enclosures as that. By June many of the lambs will be ripe, by July many of the others, and even when the lambs are born late, when managed in this way they should all be ripe as peaches by the middle of August. After the lambs are gone the ewes can be managed a little less carefully, especially if they are in strong condition, though there is a comfort in knowing that every stomach worm germ that falls to the earth must die from lack of a host.

To make this thing doubly successful put flat bottomed troughs in the pastures ahead, where the lambs run, and put feed in them; any sort of grain, corn, oats, barley, bran, coarse-ground or broken cake or oil meal. Thus the lambs will grow like weeds and pay many times over for their grain. Thus more sheep may be carried on the same ground than would be possible under ordinary treatment. There is scarcely any limit to the number of sheep that can be safely kept on an eastern farm under this system of management. The limit is, of course, the size of the farm and the amount of grass. Even this can be greatly helped by soiling. Racks may with great profit be placed in the fields and the ewes fed with green crops, fresh mown oats, peas, clover or alfalfa. Thus twice as many ewes may be kept as the grass alone will support. The writer would suggest that about 400 ewes would keep one man nicely busy in caring for them and their lambs, hauling water to them, soiling somewhat, and feeding the lambs. He would not hesitate to undertake the management of 400 ewes on one farm in any part of the corn belt, the regions most infested with stomach worms. There is no business more sure of profit than this. Lambs sell remarkably well and the prospect is that as the western ranges are diminished that they will sell better for the ravages of the stomach worm deter eastern farmers from going into the business. The two serious obstacles to be overcome are: first, the question of water and next, the question of shade. Water is readily hauled in mounted tanks as it is usually in England. Shade is not absolutely essential. The writer has seen very fat sheep in the San Joaquin valley of California confined to the alfalfa meadows and with no shade whatever. Probably a system of canvas sheds, long and narrow, would not be very expensive nor too troublesome for one man to move and set up unaided. Any sort of grass will serve. Kentucky blue grass is to be preferred, perhaps brome grass (Bromus inermis) is better, clovers may be utilized and oats sown to be grazed off, with peas.

The writer does not hesitate to say that he looks forward to seeing many sheep farms established in the corn belt each carrying from 200 to 500 ewes and managed nearly under this system. He feels confident that no other branch of the live stock industry holds forth better prospects.

It should be borne in mind that the earlier the lambs are born the sooner they will be gone to market, and thus the fewer pastures will be needed. Also the market is usually best in June and July, after the flood of fed lambs has passed and before the new crop from the ranges has started to come.

Besides the stomach worm there is the worm that makes the nodular disease of the intestines. Any observant man who has dissected a mature sheep has often noticed on the small intestines little nodules or "knots." These are really small tumors, filled with a greenish, cheesy substance. They do not do much harm when they are few in number but the trouble is a cumulative one and the numbers of the nodules increase until after a time digestion and absorption are much interfered with. Sometimes parts of the intestines become calcified, that is, so impregnated with lime salts that they are almost like stone. Death ensues in a longer or shorter time from the nodular disease. It does not work quickly as does the disease caused by the stomach worm. The worm causing these tumors is called œsophagostoma columbianum.

This nodular disease is a hard one to cure, if indeed it is possible to cure it at all after it is established. Prevention is about all that we can do. Dr. W. H. Dalrymple of the Louisiana Experiment Station has shown, however, that it is readily communicable from affected ewes to their lambs through the medium of the pasture. He has also demonstrated that where diseased ewes are kept confined to the barn and their lambs allowed to run on clean pasture not contaminated by the presence of any old sheep, the lambs remain healthy and thus a new healthful stock can be had even from a diseased flock. None of these diseases originates spontaneously. There are no other known hosts of these diseases than sheep, goats and perhaps deer, so it is merely a question of starting with the lambs, born free of all parasites, and keeping them in health by putting them on fresh and uninfected pasture.

In this little manual only the most essential things have been sought to be presented. The flock owner who wishes to know more of detail of management is advised to secure the new and complete book by Joseph E. Wing, called "Sheep Farming in America." It is published by the Sanders Publishing Co., 358 Dearborn St., Chicago, and the price is $1.00. This book is very fully packed with practical information and results of experience, not only with Dorsets but with other breeds and crossbred sheep.

The Continental Dorset Club publishes the flock register for pure-bred Dorset Horn Sheep in America. Information concerning registration, or lists of breeders in America may be had of the secretary, Joseph E. Wing, Woodland Farm, Mechanicsburg, Ohio.

FINIS